Number Theory
for Beginners

André Weil

Number Theory for Beginners

With the Collaboration of

Maxwell Rosenlicht

Springer-Verlag

New York   Heidelberg   Berlin

André Weil
Institute for Advanced Study
Princeton, NJ 08540
USA

Maxwell Rosenlicht
University of California
Berkeley, CA 94720
USA

AMS Subject Classification: 12-01

Library of Congress Cataloging in Publication Data

Weil, André, 1906–
  Number theory for beginners.

  Includes index.
  1. Numbers, Theory of. I. Rosenlicht, Maxwell,
  joint author. II. Title.
  QA241.W342   512'.7   79-10764

Printed in the United States of America.

9 8 7 6 5 4 3 2 1

ISBN 0-387-90381-x Springer-Verlag New York

ISBN 3-540-90381-x Springer-Verlag Berlin Heidelberg

FOREWORD

In the summer quarter of 1949, I taught a ten-weeks introductory course on number theory at the University of Chicago; it was announced in the catalogue as "Algebra 251". What made it possible, in the form which I had planned for it, was the fact that Max Rosenlicht, now of the University of California at Berkeley, was then my assistant. According to his recollection, "this was the first and last time, in the history of the Chicago department of mathematics, that an assistant worked for his salary". The course consisted of two lectures a week, supplemented by a weekly "laboratory period" where students were given exercises which they were asked to solve under Max's supervision and (when necessary) with his help. This idea was borrowed from the "Praktikum" of German universities. Being alien to the local tradition, it did not work out as well as I had hoped, and student attendance at the problem sessions soon became desultory.

Weekly notes were written up by Max Rosenlicht and issued week by week to the students. Rather than a literal reproduction of the course, they should be regarded as its skeleton; they were supplemented by references to standard text-books on algebra. Max also contributed by far the larger part of the exercises.

None of this was meant for publication. Now some friends of mine, having come across these notes by chance, have concluded that they could still be useful, and the Springer Verlag, ever anxious to print more and more mathematics, has persuaded me to publish them. They took very little editing; here they are, for whatever they may be worth. Experience seems to show that they should be somewhat diluted for class use. The subdivision into §§ corresponds only roughly to the lectures as given. For lack of time, the quadratic reciprocity law had to be omitted from the lectures and was only issued in written form.

October 16, 1978   *A. Weil*

CONTENTS

We assume as known the concepts of "set" and "subset"; $\in$ means "element of". We use **Z** for the set of all integers, and **Q** for the set of all rational numbers. We assume the basic properties of integers and of rational numbers:

(1) $(x+y)+z = x+(y+z)$.
(2) $x+y = y+x$.
(3) *The equation $a+x = b$ has a unique solution x* (in **Z**, if a,b are in **Z**; in **Q**, if they are in **Q**).
(4) $0+x = x$.
(1') $(xy)z = x(yz)$.
(2') $xy = yx$.
(3') *The equation $ax = b$ has a unique solution x in **Q**, if a,b are in **Q** and $a \neq 0$.*
(4') $1 \cdot x = x$.
(5) $x(y+z) = xy + xz$ (this is the "distributive law").

The unique solution of $a+x=b$ is written $b-a$. For $a\neq 0$, the unique solution of $ax=b$ is written $\dfrac{b}{a}$.

A rational number is *positive* ($\geqslant 0$) or *negative* ($\leqslant 0$); only 0 is both; $b\geqslant a$ (or $a\leqslant b$) means $b-a\geqslant 0$; $b>a$ (or $a<b$) means $b\geqslant a$, $b\neq a$. If $x>0$ and $y>0$, then $x+y>0$ and $xy>0$.

If a,b,x are integers, and $b=ax$, b is said to be a *multiple of* a; a is said to *divide* b or to be a *divisor of* b; when that is so, we write $a|b$.

Finally, we have:

(6) *Any non-empty set of positive integers contains a least integer.*

In fact, such a set contains some integer n; then the first one among the integers $0, 1, \ldots, n-1$, n to be contained in the set has the property in question. An equivalent form of (6) is the "*principle of mathematical induction*":

(6′) *If a statement about a positive integer x is true for $x=0$, and its truth for all $x<n$ implies its truth for $x=n$, then it is true for all x.*

PROOF. Call F the set consisting of the positive integers for which the statement is not true. If F is not empty, apply (6) to it; the conclusion contradicts the assumption in (6′).

EXERCISES

I.1. Show that the relation $(-1)\cdot(-1)=+1$ is a consequence of the distributive law.

I.2. Prove that any integer $x > 1$ has either a divisor > 1 and $\leqslant \sqrt{x}$ or no divisor > 1 and $< x$ (in the latter case it is called a prime; cf. § IV).

I.3. Prove by induction that

$$1^3 + 2^3 + \cdots + n^3 = \left[\frac{n(n+1)}{2} \right]^2.$$

I.4. Prove by induction that $4^{2n+1} + 3^{n+2}$ is a multiple of 13 for $n \geqslant 0$.

I.5. If one is given a balance, and n weights of $1, 3, 3^2, \ldots, 3^{n-1}$ lb. respectively, show that it is possible, by placing some weights in one pan and some in the other, to weigh out any weight of N lb., with N an integer $\geqslant 1$ and $\leqslant 1/2 \, (3^n - 1)$ (Hint: consider all sums of the form

$$e_0 + 3e_1 + 3^2 e_2 + \cdots + 3^{n-1} e_{n-1},$$

where each e_i is 0, $+1$ or -1).

I.6. Show that the number of terms in any polynomial of degree d in n variables is at most $\dfrac{(n+d)!}{n!d!}$ (Hint: use induction on d, and observe that the number of terms in a homogeneous polynomial of degree d in n variables is the same as that of a polynomial of degree d in $n-1$ variables).

Lemma. *Let d, a be integers, $d > 0$. Then there exists a unique largest multiple dq of d which is $\leqslant a$; it can be characterized by $dq \leqslant a < d(q+1)$, or also by $a = dq + r$, $0 \leqslant r < d$.*

(In that relation, r is called the remainder of the division of a by d; d is called the divisor, and q the quotient).

PROOF. The set of integers of the form $a - dz$ with $z \in \mathbf{Z}$ contains positive integers, since z can be taken negative and large in absolute value (i.e. $z = -N$, with N a large positive integer). Apply (6) of § I to the set of all positive integers which can be written in that form; take its smallest element r, and write it as $a - dq$; this is $\geqslant 0$ and $< d$, since otherwise $a - d(q+1)$ would belong to the same set and would be $< r$. $\qquad\square$

Theorem II.1. *Let M be a non-empty set of integers, closed under subtraction. Then there exists a unique $m \geqslant 0$ such that M is the set of all multiples of m: $M = \{mz\}_{z \in \mathbf{Z}} = m\mathbf{Z}$.*

PROOF. If $x \in M$, then, by assumption, $0 = x - x \in M$ and $-x = 0 - x \in M$. If also $y \in M$, then $y + x = y - (-x) \in M$, so that M is also closed under addition. If $x \in M$ and $nx \in M$, where n is any positive integer, then $(n+1)x = nx + x \in M$; therefore, by induction, $nx \in M$ for all $n \geqslant 0$, hence also for all $n \in \mathbf{Z}$. Finally, all linear combinations of elements of M with integral coefficients are in M; as this property of M obviously implies that M is closed under addition and subtraction, it is equivalent with our assumption on M.

If $M = \{0\}$, the theorem is true with $m = 0$. If not, the set of elements > 0 in M cannot be empty; take for m the smallest one. All multiples of m are then in M. For any $x \in M$, apply the lemma and write $x = my + r$ with $0 \leqslant r < m$; then $r = x - my$ is in M. In view of the definition of m, this implies $r = 0$, $x = my$. Therefore $M = m\mathbf{Z}$. Conversely, since m is the smallest element > 0 in $m\mathbf{Z}$, it is uniquely determined when M is given.

Corollary 1. *Let $a, b, \ldots, c$ be integers in any (finite) number. Then there is a unique integer $d \geqslant 0$ such that the set of all linear combinations $ax + by + \cdots + cz$ of $a, b, \ldots, c$ with integral coefficients $x, y, \ldots, z$ is the set of all multiples of d.*

PROOF. Apply theorem II.1 to that set.   □

Corollary 2. *Assumptions and notations being the same as in corollary 1, d is a divisor of each one of the integers $a, b, \ldots, c$, and every common divisor of these integers is a divisor of d.*

PROOF. Each one of the integers $a, b, \ldots, c$ belongs to the set of their linear combinations. Conversely, every common divisor of $a, b, \ldots, c$ is a divisor of every one of their linear combinations, hence in particular of d.

Definition. The integer d defined in the corollaries of theorem II.1 is called the greatest common divisor (or in short the g.c.d.) of $a, b, \ldots, c$; it is denoted by $(a, b, \ldots, c)$.

As the g.c.d. $(a, b, \ldots, c)$ belongs to the set of linear combinations of $a, b, \ldots, c$ (since it is the smallest element > 0 of that set, unless $a, b, \ldots, c$ are all 0), it can be written in the form

$$(a, b, \ldots, c) = ax_0 + by_0 + \cdots + cz_0$$

where $x_0, y_0, \ldots, z_0$ are all integers.

EXERCISES

II.1. Prove that $(a, b, c) = ((a, b), c) = (a, (b, c))$.

II.2. Prove that, in the "series of Fibonacci" $1, 2, 3, 5, 8, 13, \ldots$, in which each term after the second is the sum of the two preceding ones, every two consecutive terms have a g.c.d. equal to 1.

II.3. If p, q, r, s are integers such that $ps - qr = \pm 1$, and a, b, a', b' are integers such that

$$a' = pa + qb, \quad b' = ra + sb,$$

prove that $(a, b) = (a', b')$ (Hint: solve the last two equations for a, b).

II.4. Let a, b be two integers > 0. Put $a_0 = a$, $a_1 = b$; for $n \geq 1$, define a_{n+1} by $a_{n-1} = a_n q_n + a_{n+1}$, $0 \leq a_{n+1} < a_n$, provided $a_n > 0$. Prove that there exists $N \geq 1$ such that $a_{N+1} = 0$, and that a_N is then equal to (a, b).

II.5. Notations being as in exercise II.4, prove that a_n can be written in the form $ax + by$, with integral x, y, for all $n \geq 0$ and $\leq N$.

II.6. Use the procedure described in exercises II.4, II.5 to find (a,b) and to solve $ax + by = (a,b)$ in each one of the following cases: (i) $a = 56$, $b = 35$; (ii) $a = 309$, $b = 186$; (iii) $a = 1024$, $b = 729$.

II.7. If $a, b, \ldots, c, m$ are integers, and $m > 0$, show that
$$(ma, mb, \ldots, mc) = m \cdot (a, b, \ldots, c).$$

II.8. Prove that every rational number can be written in one and only one way as $\dfrac{m}{n}$ with $(m, n) = 1$ and $n > 0$.

§ III

Definition. Integers $a, b, \ldots, c$ are called mutually relatively prime if their g.c.d. is 1.

In other words, they are mutually relatively prime if they have no other common positive divisor than 1.

If two integers a, b are mutually relatively prime, then a is said to be prime to b, and b prime to a; when that is so, every divisor of a is also prime to b, and every divisor of b is prime to a.

Theorem III.1. *Integers $a, b, \ldots, c$ are mutually relatively prime if and only if the equation $ax + by + \cdots + cz = 1$ has a solution in integers $x, y, \ldots, z$.*

In fact, if $(a, b, \ldots, c) = 1$, then, by corollary 1 of theorem II.1, the equation in question has a solution. Conversely, if it has a solution, then every common divisor $d > 0$ of $a, b, \ldots, c$ must divide 1 and must therefore be 1.

9

Corollary. *If d is the g.c.d. of integers $a, b, \ldots, c$, then $\dfrac{a}{d}, \dfrac{b}{d}, \ldots, \dfrac{c}{d}$ are mutually relatively prime.*

This follows at once from the fact that d can be written in the form $ax_0 + by_0 + \cdots + cz_0$.

Theorem III.2. *If a, b, c are integers such that a is prime to b and divides bc, then a divides c.*

As $(a, b) = 1$, we can solve $ax + by = 1$. Then we have

$$c = c(ax + by) = a(cx) + (bc)y.$$

As a divides both terms in the right-hand side, it divides c.

Corollary 1. *If a, b, c are integers, and if a is prime both to b and to c, it is prime to bc.*

If d is a positive common divisor of a and of bc, it is prime to b (since it divides a) and must therefore divide c, by theorem III.2. As $(a, c) = 1$, d must be 1.

Corollary 2. *If an integer is prime to each one of the integers $a, b, \ldots, c$, it is prime to their product.*

This follows from corollary 1 by induction on the number of factors in that product.

EXERCISES

III.1. If $(a, b) = 1$ and both a and b divide c, show that ab divides c.

III.2. If $m > 1$ and a is prime to m, show that the remainders obtained by dividing $a, 2a, \ldots, (m-1)a$ by m (cf. § II, lemma) are the numbers $1, 2, \ldots, m-1$, in some order.

III.3. Show that, if N is an integer > 0, either it is a "perfect square" (i.e. of the form n^2, where n is an integer > 0) or $\sqrt{N}$ is not a rational number (Hint: use exercise II.8).

III.4. Any integer > 1 which is not a power of 2 can be written as the sum of (two or more) consecutive integers.

III.5. If a, b are positive integers, and $(a, b) = 1$, show that every integer $\geqslant ab$ can be written in the form $ax + by$ with positive integers x, y.

III.6. Using exercise III.5 and induction on m, show that, if $a_1, a_2, \ldots, a_m$ are positive integers and $d = (a_1, a_2, \ldots, a_m)$, every sufficiently large multiple of d can be written in the form $a_1 x_1 + a_2 x_2 + \cdots + a_m x_m$, where the x_i are positive integers.

Definition. An integer $p > 1$ is called a prime if it has no other positive divisor than itself and 1.

Every integer > 1 has at least one prime divisor, viz., its smallest divisor > 1. If a is any integer, and p is a prime, then either p divides a or it is prime to a.

Theorem IV.1. *If a prime divides the product of some integers, it divides at least one of the factors.*

For otherwise it would be prime to all those factors, and therefore, by corollary 2 of theorem III.2, it would be prime to their product.

Theorem IV.2. *Every integer > 1 can be written as a product of primes; it can be so written in only one way except for the order of the factors.*

Take $a>1$; call p a prime divisor of a. If $a=p$, the theorem holds for a; if not, $\frac{a}{p}$ is >1 and $<a$; if the first statement in the theorem holds for $\frac{a}{p}$, it holds for a. Therefore that statement follows, by induction on a.

The second statement can also be proved by induction, as follows. Assume that a is written in two different ways as a product of primes, say as $a=pq\ldots r$ and as $a=p'q'\ldots s'$; as p divides a, theorem IV.1 implies that p must divide one of the primes $p',q',\ldots,s'$, say p'. Then $p=p'$; applying the second part of the theorem to $\frac{a}{p}$, we see that $q',\ldots,s'$ must be the same as $q,\ldots,r$, except for their order. By induction, this proves the second part.

It is worth while to give another proof. Write a as a product of primes, $a=pq\ldots r$; let P be any prime; let n be the number of times that P occurs among the factors $p,q,\ldots,r$ of a. Then a is a multiple of P^n; on the other hand, since $a\cdot P^{-n}$ is a product of primes other than P, it is not a multiple of P, by theorem IV.1, and therefore a is not a multiple of P^{n+1}. Thus n is uniquely determined as the largest integer such that P^n divides a; we can write $n=v_P(a)$. Thus, for any two ways of writing a as a product of primes, the same primes must occur, and must occur the same number of times in both products; this again proves the second part of our theorem.

2 is a prime; it is the only even prime; all others are odd. The first ten primes are

$$2, 3, 5, 7, 11, 13, 17, 19, 23, 29.$$

Let $p_1=2,p_2,p_3,\ldots$ be all the primes in their natural (i.e. increasing) order. Let a be any integer $\geqslant 1$; for each $i\geqslant 1$, let α_i be the number of times that p_i occurs among the prime factors of a when a is written as a product of such factors (with $\alpha_i = 0$ if p_i does not divide a). Then we have

$$a = p_1^{\alpha_1} p_2^{\alpha_2} \ldots p_r^{\alpha_r}$$

provided r has been taken large enough (i.e. so large that all distinct prime divisors of a occur among $p_1, p_2, \ldots, p_r$).

Theorem IV.3. *There are infinitely many primes.*

In fact, if $p, q, \ldots, r$ are primes in any finite number, any prime divisor of $pq \ldots r + 1$ must be distinct from $p, q, \ldots, r$. (This is Euclid's proof for that theorem; for other proofs, cf. the exercises).

EXERCISES

IV.1. Let n be an integer $\geqslant 1$, and p a prime. If for any rational number x we denote by $[x]$ the largest integer $\leqslant x$, show that the largest integer N such that p^N divides $n!$ is given by

$$N = \left[\frac{n}{p}\right] + \left[\frac{n}{p^2}\right] + \left[\frac{n}{p^3}\right] + \cdots$$

IV.2. Prove that, if $a, b, \ldots, c$ are integers $\geqslant 1$, then
$$\frac{(a + b + \cdots + c)!}{a! b! \ldots c!}$$
is an integer. (Hint: using ex. IV.1, show that every prime occurs at least as many times in the numerator as in the denominator).

IV.3. If a, m, n are positive integers, and $m \neq n$, show that the g.c.d. of $a^{2^m} + 1$ and $a^{2^n} + 1$ is 1 or 2 according as a is even or odd (Hint: use the fact that $a^{2^n} - 1$ is a multiple of $a^{2^m} + 1$ for $n > m$). From this, deduce the existence of infinitely many primes.

IV.4. If $a = p^\alpha q^\beta \ldots r^\gamma$, where $p, q, \ldots, r$ are distinct primes and $\alpha, \beta, \ldots, \gamma$ are positive integers, prove that the number of distinct divisors of a, including a and 1, is

$$(\alpha + 1)(\beta + 1) \ldots (\gamma + 1)$$

and that their sum is

$$\frac{p^{\alpha + 1} - 1}{p - 1} \cdot \frac{q^{\beta + 1} - 1}{q - 1} \ldots \frac{r^{\gamma + 1} - 1}{r - 1}.$$

IV.5. Prove that, if D is the number of distinct divisors of a, their product is $a^{D/2}$.

IV.6. If $n, a, b, \ldots, c$ are integers > 1, then the number of distinct integers of the form $a^\alpha b^\beta \ldots c^\gamma$ which are $\leqslant n$ (where $\alpha, \beta, \ldots, \gamma$ are positive integers) is

$$\leqslant \left(1 + \frac{\log n}{\log a}\right)\left(1 + \frac{\log n}{\log b}\right) \cdots \left(1 + \frac{\log n}{\log c}\right).$$

Using this fact, and the fact that

$$\lim_{n \to +\infty} \frac{(\log n)^r}{n} = 0$$

for every $r > 0$, prove that the number of primes is infinite (Hint: assuming it to be finite, take for $a, b, \ldots, c$ all the distinct primes).

IV.7. If $(a, b) = 1$ and $a^2 - b^2$ is a "perfect square" (cf. ex. III.3), show that either $a + b$ and $a - b$ are perfect squares, or each is twice a perfect square (Hint: show that the g.c.d. of $a + b$ and $a - b$ is 1 or 2).

Definition. A commutative (or "abelian") group is a set G, together with a binary operation between elements of G, satisfying the following axioms (in which the group operation is denoted by $+$):

I *(Associativity)*. $(x+y)+z=x+(y+z)$ for all x,y,z in G.

II *(Commutativity)*. $x+y=y+x$ for all x,y in G.

III If x,y are in G, the equation $x+z=y$ has a unique solution z in G (written $z=y-x$).

IV There is an element in G, called the neutral element (and denoted by 0) such that $0+x=x$ for all x in G.

For instance, the integers, the rational numbers (and the real numbers) are commutative groups under the operation of addition. There are of course many cases of commutative groups where the group operation is not written additively, i.e. is not denoted by $+$; then the

17

notations $y - x$ in III, and 0 in IV, are to be suitably modified. If the operation is written as a multiplication, then one usually writes $\frac{y}{x}$, or y/x, or yx^{-1}, for the element z in III, and 1 for the neutral element in IV. Non-zero rational numbers give an example of a commutative group under multiplication.

In the present book, no groups will occur, other than commutative groups; therefore the word "commutative" will mostly be omitted. A subset of a group G which makes up a group under the same binary operation as G is called a *subgroup* of G. If G is written additively, a subset of G is a subgroup if and only if it is closed under addition and subtraction, or even merely under subtraction (cf. the proof of theorem II.1). Theorem II.1 may be expressed concisely by saying that every subgroup of $\mathbf{Z}$ is of the form $m\mathbf{Z}$ with $m \geqslant 0$.

Examples will now be given of finite groups.

Definition. If m, x, y are integers and $m > 0$, x and y are said to be congruent modulo m if $x - y$ is a multiple of m; then one writes $x \equiv y \pmod{m}$, or more briefly $x \equiv y \ (m)$.

The lemma in § II shows that every integer is congruent modulo m to one and only one of the integers $0, 1, \ldots, m - 1$, and that two integers are congruent modulo m if and only if they have the same remainder in the division by m.

The relation of congruence modulo m has the following properties:

(A) (*Reflexivity*) $x \equiv x \pmod{m}$;
(B) (*Transitivity*) $x \equiv y$ and $y \equiv z \pmod{m}$ implies $x \equiv z$ $\pmod{m}$;
(C) (*Symmetry*) $x \equiv y \pmod{m}$ implies $y \equiv x \pmod{m}$.

(D) $x \equiv y$, $x' \equiv y'$ (mod m) implies $x \pm x' \equiv y \pm y'$ (mod m).

(E) $x \equiv y$, $x' \equiv y'$ (mod m) implies $xx' \equiv yy'$ (mod m).

(F) Let $d > 0$ divide m, x and y; then $x \equiv y$ (mod m) if and only if $\dfrac{x}{d} \equiv \dfrac{y}{d}$ $\left(\text{mod } \dfrac{m}{d} \right)$.

As to (E), it is a consequence of the identity

$$xx' - yy' = x(x' - y') + (x - y)y';$$

verification of all the other statements is immediate.

Properties (A), (B), (C) are expressed by saying that the relation of congruence modulo m is an "equivalence relation" between integers.

Definition. A congruence class of integers modulo m is a set consisting of all the integers which are congruent to a given one modulo m.

If x is any integer, we write (x mod m) (or simply (x) if no ambiguity can occur) for the congruence class of the integers congruent to x modulo m. By (A), x belongs to the class (x mod m); it is called a *representative* of that class. From (A), (B), (C), it follows that any two classes (x mod m), (y mod m) coincide if $x \equiv y$ (mod m) and are disjoint (i.e., have no element in common) otherwise. Thus the set of all integers is separated into m disjoint classes (0 mod m), (1 mod m),...,($m - 1$ mod m).

We define the addition of congruence classes by putting:

$$(x \bmod m) + (y \bmod m) = (x + y \bmod m);$$

this is meaningful, for (D) shows that the right-hand side depends only upon the two classes in the left-hand side and not upon the choice of the representatives x,y for these classes.

Theorem V.1. *For any integer $m>0$, the congruence classes modulo m, under addition, make up a commutative group of m elements.*
    It is immediate, in fact, that the equation

$$(x \bmod m)+(z \bmod m)=(y \bmod m),$$

for given x,y, has the unique solution $(y-x \bmod m)$, and that $(0 \bmod m)$ has the property required of the neutral element.

EXERCISES

V.1. If $x_1,\ldots,x_m$ are m integers, show that the sum of a suitable non-empty subset of that set is a multiple of m (Hint: consider the distinct classes modulo m among those determined by $0,x_1,x_1+x_2,\ldots,x_1+x_2+\cdots+x_m$).

V.2. Prove that every "perfect square" (cf. ex. III.3) is congruent to 0, 1 or 4 modulo 8.

V.3. Prove by induction that, if n is a positive integer, then
$$2^{2n+1}\equiv 9n^2-3n+2 \ (\bmod\ 54).$$

V.4. Show that, if x,y,z are integers, and $x^2+y^2=z^2$, then $xyz\equiv 0 \ (\bmod\ 60)$.

V.5. If $x_0,x_1,\ldots,x_n$ are integers, show that
$$x_0+10x_1+\cdots+10^n x_n\equiv x_0+x_1+\cdots+x_n \ (\bmod\ 9).$$

V.6. Show that a necessary and sufficient condition for the pair of congruences $x\equiv a \ (\bmod\ m)$, $x\equiv b \ (\bmod\ n)$ to have a solution is that $a\equiv b \ (\bmod\ d)$, where $d=(m,n)$. If $d=1$, show that the solution is unique modulo mn.

V.7. If n is an integer >0, show that any $n+1$ of the first $2n$ integers contain a pair x,y such that $\dfrac{y}{x}$ is a power of 2 (Hint: for each one of the given integers $x_0, x_1, \ldots, x_n$, call x_i' the largest odd divisor of x_i, and show that at least two of these must be equal).

V.8. When x,y are two integers >0, write $x \sim y$ if $\dfrac{y}{x}$ is a power of 2, i.e. of the form 2^n with $n \in \mathbf{Z}$; show that this is an equivalence relation, and that $x \sim y$ if and only if the odd divisors of x are the same as those of y.

§ VI

If m is any integer > 0, we define the multiplication of congruence classes by putting

$$(x \bmod m) \cdot (y \bmod m) = (xy \bmod m);$$

in fact, property (E), § V, shows that the right-hand side depends only upon the two classes in the left-hand side and not upon the choice of their representatives x, y.

Definition. A ring is a set R, together with two binary operations, an addition (written $+$) and a multiplication (written $\cdot$ or $\times$) satisfying the following axioms:

I. Under addition, R is a group.

II. Multiplication is associative, commutative and distributive with respect to addition: $(xy)z = x(yz)$, $xy = yx$, $x(y+z) = xy + xz$ for all x, y, z.

If R is a ring, then, by the distributive law

$$(x \cdot 0) + (xz) = x(0 + z) = xz,$$

so that $x \cdot 0 = 0$ by the additive group property. Similarly, $x \cdot (-y) = -xy$.

If there is in R an element e such that $ex = x$ for all x, this is unique; for, if f is also such, then $ef = f$ and $ef = fe = e$. Such an element is called *the unit element* and is frequently denoted by 1_R or by 1; a ring is called *unitary* if it has a unit element.

The set $\mathbf{Z}$ of the integers, and the set $\mathbf{Q}$ of the rational numbers, are unitary rings.

Theorem VI.1. *For any integer $m > 0$, the congruence classes modulo m, under addition and multiplication, make up a unitary ring of m elements.*

The verification is immediate. The unit element is the congruence class $(1 \bmod m)$; for that class, we will usually write 1, and 0 for the class $(0 \bmod m)$; we have $1 \neq 0$ unless $m = 1$. The example $m = 6$ shows that, in a unitary ring, xy may be 0 without either x or y being 0 (take for x, y the classes of 2 and of 3 modulo 6); when that is so, x and y are called *zero-divisors*. The rings $\mathbf{Z}, \mathbf{Q}$ are without zero-divisors.

If a is prime to m, and $a' = a + mt$, then every common divisor of a' and m must also divide $a = a' - mt$; this shows that all integers in the congruence class $(a \bmod m)$ are then prime to m. Such a class will be called *prime to m*. If $(a \bmod m)$, $(b \bmod m)$ are both prime to m, so is $(ab \bmod m)$, by corollary 1 of theorem III.2; in particular, such classes cannot be zero-divisors in the ring of congruence classes modulo m.

Theorem VI.2. *Let* m, a, b *be integers, with* $m > 0$; *put* $d = (a, m)$. *Then the congruence* $ax \equiv b \pmod{m}$ *has either exactly* d *solutions modulo* m, *or no solution; it has a solution if and only if* $b \equiv 0 \pmod{d}$; *there are exactly* $\dfrac{m}{d}$ *distinct values of* b *modulo* m *for which this is so.*

In fact, x is a solution if and only if there is an integer y such that $ax - b = my$, i.e. $b = ax - my$; by corollary 1 of theorem II.1, this has a solution if and only if d divides b, i.e. $b = dz$; we get all distinct values of b modulo m of that form by giving to z the values $0, 1, \ldots, \dfrac{m}{d} - 1$. If x is a solution of $ax \equiv b \pmod{m}$, then x' is also a solution if and only if $a(x' - x) \equiv 0 \pmod{m}$; by property (F) of congruences, this is equivalent to $\dfrac{a}{d}(x' - x) \equiv 0 \left(\bmod \dfrac{m}{d} \right)$, and therefore to $x' \equiv x \left(\bmod \dfrac{m}{d} \right)$ by theorem III.2 and the corollary of theorem III.1. This shows that all the solutions of $ax' \equiv b \pmod{m}$ can be written as $x' = x + \dfrac{m}{d} u$; the distinct solutions modulo m are then obtained by giving to u the values $0, 1, \ldots, d - 1$.

Corollary. *The congruence classes prime to* m *modulo* m *make up a group under multiplication.*

This follows at once from corollary 1 of theorem III.2, from theorem VI.2, and from the fact that the class $(1 \bmod n)$ is the neutral element for multiplication in the ring of congruence classes modulo m.

Definition. For any integer $m > 0$, the number of congruence classes prime to m modulo m is denoted by $\varphi(m)$, and φ is called the Euler function.

Accordingly, we have

$$\varphi(1)=\varphi(2)=1, \quad \varphi(3)=\varphi(4)=2, \quad \varphi(5)=4, \text{ etc.}$$

If $m \geqslant 2$, $\varphi(m)$ can also be defined as the number of positive integers prime to m and $\leqslant m-1$. If p is a prime, $\varphi(p)=p-1$.

Definition. A field is a ring whose non-zero elements make up a group under multiplication.

The ring **Q** of rational numbers is a field; the ring **Z** of integers is not a field. A field has no zero-divisors; the example of **Z** shows that the converse is not true.

Theorem VI.3. *For any integer $m > 1$, the ring of congruence classes modulo m is a field if and only if m is a prime.*

If m is a prime, all classes modulo m, other than 0, are prime to m and therefore make up a multiplicative group, by the corollary of theorem VI.2. On the other hand, if m is not a prime, it has a divisor d which is >1 and $<m$; then $1 < \dfrac{m}{d} < m$, so that the classes $(d \bmod m)$, $\left(\dfrac{m}{d} \bmod m\right)$ are not 0 while their product is 0. Therefore they are zero-divisors, and the ring modulo m is not a field.

If p is a prime, the field of congruence classes modulo p will be denoted by $\mathbf{F}_p$; it has p elements.

EXERCISES

VI.1. If $F(X)$ is a polynomial with integral coefficients, and if $x \equiv y \pmod{m}$, show that $F(x) \equiv F(y) \pmod{m}$.

VI.2. Solve the pair of congruences

$$5x - 7y \equiv 9 \ (\text{mod } 12), \quad 2x + 3y \equiv 10 \ (\text{mod } 12);$$

show that the solution is unique modulo 12.

VI.3. For all values of a and b modulo 2, solve

$$x^2 + ax + b \equiv 0 \ (\text{mod } 2).$$

VI.4. Solve $x^2 - 3x + 3 \equiv 0 \ (\text{mod } 7)$.

VI.5. If $m > 1$, show that the arithmetic mean of all positive integers prime to m and $< m$ is $\dfrac{m}{2}$.

VI.6. If m is any odd integer, prove that

$$1^m + 2^m + \cdots + (m-1)^m \equiv 0 \ (\text{mod } m).$$

VI.7. If m, n are integers > 0, and $(m, n) = 1$, prove that $\varphi(mn) = \varphi(m)\varphi(n)$ (Hint: use exercise V.6).

VI.8. Show that, if $m > 1$ and $p, q, \ldots, r$ are all the distinct prime divisors of m, then

$$\varphi(m) = m\left(1 - \frac{1}{p}\right)\left(1 - \frac{1}{q}\right) \cdots \left(1 - \frac{1}{r}\right).$$

VI.9. If p is any prime, prove by induction on n, using the binomial formula, that $n^p \equiv n \ (\text{mod } p)$ for all integers n.

VI.10. If p is any prime, and $n \geqslant 0$, prove by induction on n that, if $a \equiv b \ (\text{mod } p)$, then $a^{p^n} \equiv b^{p^n} \ (\text{mod } p^{n+1})$.

VI.11. If p is an odd prime, and $x^2 \equiv y^2 \ (\text{mod } p)$, show that x is congruent either to y or to $-y$ modulo p, but not to both unless p divides x and y; hence conclude that $x^2 \equiv a \ (\text{mod } p)$ has a solution x for exactly half of the integers a among $1, 2, \ldots, p - 1$.

VI.12. Show that the numbers of the form $x + y\sqrt{2}$, where x and y are integers, make up a ring; if x, y range over all rational numbers, show that they make up a field.

§ VII

The definition of a group and of a subgroup makes it clear that the intersection of subgroups of a group G (in any number, finite or not) is again a subgroup of G.

Definition. Let $a, b, \ldots, c$ be elements of a group G. Then the intersection G' of all the subgroups of G (including G itself) which contain $a, b, \ldots, c$ is said to be generated by $a, b, \ldots, c$, and these are said to be generators of G'.

This can also be expressed by saying that G' is the smallest subgroup of G containing $a, b, \ldots, c$; it may happen that it is no other than G itself.

Let G be a group, x an element of G, and call G_x the subgroup generated by x. Suppose that G is written additively. As usual, write $-x$ for $0 - x$; this must be in G_x. Also, write $0 \cdot x$ for 0; for every integer $n > 0$, write nx for the sum $x + x + \cdots + x$ of n terms all equal to x, and $(-n)x$ for $-(nx)$; by induction on n, all these must be in

G_x. Also by induction, one verifies at once the formulas

$$mx + nx = (m+n)x, \quad m(nx) = (mn)x.$$

for all m,n in $\mathbf{Z}$. The first formula shows that the elements nx, for $n \in \mathbf{Z}$, make up a subgroup of G; clearly, this is no other than G_x. For convenience, we state this as a theorem only in the case when G is written multiplicatively; then we write x^0 for the neutral element 1 of G, x^{-1} for the element x' defined by $x'x = 1$, x^n for the product $x \cdot x \cdot \ldots \cdot x$ of n factors equal to x, and x^{-n} for $(x^n)^{-1}$.

Theorem VII.1. *Let G be a group under multiplication; then, for every $x \in G$, the subgroup of G generated by x consists of the elements x^n for $n \in \mathbf{Z}$.*

G and x being as in theorem VII.1, call M_x the set of those integers a for which $x^a = 1$. As $x^0 = 1$, M_x is not empty. Also we have, for all integers a,b:

$$x^{a-b} = x^a \cdot (x^b)^{-1},$$

which shows that $x^a = x^b$ if and only if $a - b$ is in M_x; in particular, M_x is closed under subtraction. Therefore M_x satisfies the assumptions in theorem II.1 (in other words, it is a subgroup of the additive group $\mathbf{Z}$) and consists of the multiples of some integer $m \geqslant 0$; if m is not 0, it is the smallest integer > 0 such that $x^m = 1$. Thus, if $m = 0$, all the elements x^a are different; if $m > 0$, x^a is the same as x^b if and only if $a \equiv b \pmod{m}$.

Definition. An isomorphism between two groups G, G' is a one-to-one correspondence (a "bijection") between the elements of G and those of G', transforming the group operation in G into the group operation in G'.

When there is such a correspondence, G and G' are said to be *isomorphic*. The concept of isomorphism can be transported in an obvious manner to rings and fields.

With this definition, the results obtained above can be reformulated as follows:

Theorem VII.2. *Let G be a group under multiplication, generated by a single element x. Then either G is infinite, and the mapping $x^a \to a$ is an isomorphism of G onto the additive group $\mathbf{Z}$, or it consists of a finite number m of elements, and then the mapping $x^a \to (a \bmod m)$ is an isomorphism of G onto the additive group of congruence classes modulo m in $\mathbf{Z}$.*

Of course, if G is any group and x any element of G, theorem VII.2 can be applied to the subgroup of G generated by x.

Definition. The number of elements of a finite group is called its *order*. If a group of finite order is generated by a single element, it is called *cyclic*; if an element x of a group generates a group of finite order m, m is also called the order of x.

EXERCISES

VII.1. If F is a finite field, show that the subgroup of the additive group of F generated by 1 is of prime order p and is a subfield of F, isomorphic to the field $\mathbf{F}_p$ of congruence classes modulo p.

VII.2. Show that a non-empty finite subset S of a group G is a subgroup of G if and only if it is closed under the group operation (Hint: if $a \in S$, then $a \to ax$ is a bijection of S onto itself).

VII.3. Show that a finite ring is a field if and only if it has no zero-divisors.

VII.4. If G is a (commutative) group, and n any integer, show that the elements x^n, for $x \in G$, make up a subgroup of G.

VII.5. If G', G'' are subgroups of the (commutative) group G, show that the elements $x'x''$ with $x' \in G', x'' \in G''$, make up a subgroup of G.

VII.6. Let G be a (commutative) group, x an element of G of order m, and y an element of G of order n. Show that, if $(m,n)=1$, then $x^a y^b = 1$ if and only if $x^a = y^b = 1$: hence show that the group generated by x and y is cyclic of order mn, generated by xy.

VII.7. Show that, if $m>2$, $n>2$, and $(m,n)=1$, the multiplicative group of congruence classes prime to mn modulo mn is not cyclic (Hint: use exercise V.6 and the fact that a cyclic group has at most one subgroup of order 2).

VII.8. Find all the values of n for which the multiplicative group of odd congruence classes modulo 2^n is cyclic.

VII.9. Show that, if G is a (commutative) group and n an integer >0, the elements of G whose order divides n make up a subgroup of G.

VII.10. Show that, if G is a finite (commutative) group, the product of all elements of G is 1 or an element of order 2.

VII.11. If p is a prime, show that $(p-1)! \equiv -1 \pmod{p}$ (Hint: consider the multiplicative group modulo p, and reason as in ex. VII.10).

§ VIII

Theorem II.1 shows that every subgroup M of $\mathbf{Z}$ is either 0 or generated by its smallest element $m > 0$; in the latter case it is generated by m or also by $-m$, but by no other element of M. For cyclic groups, we have:

Theorem VIII.1. *Let G be a cyclic group of order m, generated by an element x. Let G' be a subgroup of G; then there is a divisor d of m such that G' is the cyclic group of order $\dfrac{m}{d}$ generated by x^d.*

Let M be the set of those integers a for which $x^a \in G'$. The formula $x^{a-b} = (x^a) \cdot (x^b)^{-1}$ shows that M is a subgroup of $\mathbf{Z}$; as it contains m, it consists of the multiples of some divisor d of m. Therefore G' consists of the elements x^{da} with $a \in \mathbf{Z}$, i.e., it is generated by x^d. We have $x^{da} = x^{db}$ if and only if $da \equiv db \pmod{m}$; by property (F) of the congruence relation (§ V), this is equivalent to $a \equiv b \left(\bmod \dfrac{m}{d} \right)$.

Corollary 1. *For every positive divisor n of m, a cyclic group of order m has one and only one subgroup of order n.*

Let G be as in theorem VIII.1, and put $d = \dfrac{m}{n}$; by that theorem, if G' is a subgroup of G of order n, it must be the one generated by x^d, and x^d does generate such a subgroup.

Corollary 2. *G, m, x, G' being as in theorem VIII.1, an element x^a of G generates G' if and only if $(a, m) = d$.*

If $(a, m) = d$, x^a is in G'; moreover, by theorem VI.2, we can solve $at \equiv d \pmod{m}$, and then we have $x^d = (x^a)^t$, so that the group generated by x^a contains x^d and hence G'.

Corollary 3. *G, m, x being as above, x^a generates G if and only if $(a, m) = 1$, and G has exactly $\varphi(m)$ distinct generators.*

Corollary 4. *For every integer $m > 0$, we have*

$$\sum_{d \mid m} \varphi(d) = m.$$

(Here the sum in the left-hand side is taken over all positive divisors d of m).

Consider a cyclic group G of order m (e.g. the additive group of congruence classes modulo m). By corollary 1, for every divisor d of m, G has exactly one subgroup G_d of order d, and $d \rightarrow G_d$ is a one-to-one correspondence between the divisors of m and the subgroups of G. For each d, call H_d the set of all distinct generators of G_d, whose number is $\varphi(d)$ by corollary 3. Each element of G belongs

to one and only one set H_d, since it generates one and
only one subgroup of G.

Let G be a group, and X a subset of G; for every $a \in G$,
we write aX for the set of the elements ax with $x \in X$. The
definition of a group implies that $x \rightarrow ax$ is a bijection of
X onto aX, so that, if X is finite, all sets aX have the same
number of elements as X.

Definition. If G is a group and H a subgroup of G, every
set of the form xH with $x \in G$ is called a coset of H in G.

Lemma. *Let xH, yH be two cosets of a subgroup H of a
group G; then, either they have no element in common, or
$xH = yH$.*

If they have a common element, this can be written as
xh with $h \in H$, and also as yh' with $h' \in H$. This
gives $y^{-1}x = h'h^{-1} \in H$, and hence $xH = y \cdot (y^{-1}x)H =
y \cdot (h'h^{-1}H) = yH$.

Theorem VIII.2. *If H is a subgroup of a finite group G, the
order of H divides the order of G.*

In fact, every element x of G belongs to some coset of
H (viz., to xH), and, by the lemma, only to one. As the
number of elements of each coset is equal to the order of
H, the order of G must be a multiple of that number.

Corollary. *If x is any element of a group of order m, its
order divides m, and $x^m = 1$.*

As the order d of x is the order of the subgroup of G
generated by x, theorem VIII.2 shows that it divides m.
Then $x^m = (x^d)^{m/d} = 1$.

(N.B. The above results, and their proofs, are valid also for other than commutative groups; as mentioned before, these remain outside the scope of our treatment).

Theorem VIII.3. *If m is any integer >0, and x an integer prime to m, then $x^{\varphi(m)} \equiv 1$ (mod m).*

This is the special case of the above corollary, when it is applied to the multiplicative group of congruence classes prime to m modulo m (or, as one says more briefly but less accurately, to the multiplicative group modulo m).

Corollary. *If p is a prime, then $x^{p-1} \equiv 1$ (mod p) for every x prime to p, and $x^p \equiv x$ (mod p) for every x.*

The first assertion is a special case of theorem VIII.3, and the second one is an immediate consequence. Conversely, the latter also implies the former, in view of theorem VI.2 (or of theorem III.2). For another proof of the second assertion, cf. exercise VI.9.

The corollary was known to Fermat and is known as Fermat's theorem; a proof was first published by Euler, who also gave a proof (substantially the same as the one given above) for theorem VIII.3, which is known as Euler's theorem.

EXERCISES

VIII.1. If G is a group of order m, and if n is prime to m, show that every element of G can be written in the form x^n with some $x \in G$.

VIII.2. If p is a prime, show that every group of order p^n, with $n>0$, contains an element of order p, and that every group of order p is cyclic.

VIII.3. If p is an odd prime divisor of $a^{2^n}+1$, with $n \geqslant 1$, show that $p \equiv 1 \pmod{2^{n+1}}$ (Hint: find the order of $(a \bmod p)$ in the multiplicative group modulo p) (N.B. This was used by Euler to show that $2^{32}+1$ is not a prime, contradicting Fermat's guess that all integers $2^{2^n}+1$ are prime).

VIII.4. If a, b are integers > 0, and $a = 2^\alpha 5^\beta m$, with m prime to 10, show that the decimal fraction for $\dfrac{b}{a}$ has a period the number of whose digits divides $\varphi(m)$; show that, if it has no period with less than $m-1$ digits, then m is a prime.

In order to consider polynomials with coefficients in a field $\mathbf{F}_p$, and equations over such fields, we begin by reviewing some elementary properties of polynomials over an arbitrary field K; these are independent of the nature of that field, and quite analogous to the properties of integers described above in §§ II, III, IV.

In this §, a field K (the "groundfield") is to be regarded as fixed once for all. A polynomial P over K (i.e. with coefficients in K), in one indeterminate X, is given by an expression

$$P(X) = a_0 + a_1 X + \cdots + a_n X^n$$

with $a_0, a_1, \ldots, a_n$ in K. If $a_n \neq 0$, P is said to be of degree n, and we write $n = \deg(P)$; every polynomial except 0 has a degree. Addition and multiplication being defined in the usual manner, such polynomials make up a ring, usually written $K[X]$. If P, Q are non-zero polynomials, then $\deg(PQ) = \deg(P) + \deg(Q)$.

Lemma. *Let A, B be two polynomials, with $B \neq 0$; put $m = \deg(B)$. Then there is a unique polynomial Q such that $A - BQ$ is 0 or of degree $<m$.*

(This should be compared with the lemma in § II). If $A = 0$, there is nothing to prove; we proceed by induction on $n = \deg(A)$: first we prove the existence of Q. If $n < m$, we take $Q = 0$. Otherwise, call bX^m the term of degree m in B, and aX^n the term of degree n in A; as the polynomial $A' = A - B \cdot \left(\dfrac{a}{b} X^{n-m} \right)$ is of degree $<n$, we can write it as $BQ' + R$ with $R = 0$ or of degree $<m$, by the induction assumption. Then $A = BQ + R$, with $Q = Q' + \dfrac{a}{b} X^{n-m}$. As to the unicity of Q, let $A - BQ$ and $A - BQ_1$ be 0 or of degree $<m$; then the same is true of $B(Q - Q_1)$; since this is of degree $m + \deg(Q - Q_1)$ unless $Q - Q_1$ is 0, Q must be the same as Q_1. □

If $R = 0$, $A = BQ$, A is said to be a *multiple of B*, and B a *divisor of A*. If $B = X - a$, R must be 0 or of degree 0, i.e. a "constant" (an element of K), so that we can write

$$A = (X - a)Q + r$$

with $r \in K$. Substituting a for X in both sides, we get $A(a) = r$; if this is 0, a is called a *root* of A. Thus A is a multiple of $X - a$ if and only if a is a root of A.

Just as theorem II.1 was derived from the lemma in § II, we have:

Theorem IX.1. *Let $\mathfrak{M}$ be a non-empty set of polynomials (over K), closed under addition and such that, if A is in $\mathfrak{M}$, all multiples of A are in $\mathfrak{M}$. Then $\mathfrak{M}$ consists of all the multiples of some polynomial D, uniquely determined up to multiplication by a non-zero constant.*

If $\mathfrak{M} = \{0\}$, the theorem is true with $D = 0$. Otherwise take in $\mathfrak{M}$ a polynomial $D \neq 0$ of smallest degree d. If A is in $\mathfrak{M}$, we can apply the lemma to A and D and write $A = DQ + R$, where R is 0 or of degree $<d$. Then $R = A + D \cdot (-Q)$ is in $\mathfrak{M}$, hence 0 by the definition of D, and $A = DQ$. If D_1 has the same property as D, then it is a multiple of D and D is a multiple of D_1, so that they have the same degree; writing then $D_1 = DE$, we see that E is of degree 0, i.e. a non-zero constant.

Call aX^d the term of degree d in D; among the polynomials differing from D only by a non-zero constant factor, there is one and only one with the highest coefficient 1, viz., $a^{-1}D$; such a polynomial will be called *normalized*.

Just as in § II, we can apply theorem IX.1 to the set $\mathfrak{M}$ of all linear combinations $AP + BQ + \cdots + CR$ of any number of given polynomials $A, B, \ldots, C$; here $P, Q, \ldots, R$ denote arbitrary polynomials. If then $\mathfrak{M}$ consists of the multiples of D, where D is either 0 or a normalized polynomial, D will be called the g.c.d. of $A, B, \ldots, C$ and will be denoted by $(A, B, \ldots, C)$. As in § II, D is a divisor of $A, B, \ldots, C$, and every common divisor of $A, B, \ldots, C$ divides D. If $D = 1$, then $A, B, \ldots, C$ are said to be mutually relatively prime; they are so if and only if there are polynomials $P, Q, \ldots, R$ such that

$$AP + BQ + \cdots + CR = 1.$$

If $(A, B) = 1$, A is said to be *prime to* B, and B to A.

A polynomial A of degree $n > 0$ is said to be *prime*, or *irreducible*, if it has no divisor of degree >0 and $<n$. Every polynomial of degree 1 is irreducible. One should note that the property of a polynomial of being irreduc-

ible need not be preserved when one changes the ground-field: for instance X^2+1 is irreducible over $\mathbf{Q}$, and also over the field of real numbers, but not over the field of complex numbers, since $X^2+1=(X+i)(X-i)$.

Exactly as in § IV, we could prove now that every polynomial of degree >0 can be written, essentially uniquely, as a product of prime polynomials. All we shall need is a weaker result:

Theorem IX.2. *Let A be a polynomial of degree $n>0$ over K; this can be written, uniquely up to the order of the factors, in the form*

$$A=(X-a_1)(X-a_2)\ldots(X-a_m)Q,$$

where $0\leqslant m\leqslant n$, $a_1,a_2,\ldots,a_m$ are in K, and Q is without roots in K.

If A has no root, this is clear; otherwise we use induction on n. If A has a root a, write $A=(X-a)A'$; as A' is of degree $n-1$, we can apply the theorem to it; writing A' in the prescribed form, we get a similar product for A. If A can be written as above and also as

$$A=(X-b_1)(X-b_2)\ldots(X-b_r)R$$

where R has no root in K, then the root a of A must occur among the a_i and also among the b_j, and then, dividing by $X-a$, we get for A' two products which, by the induction assumption, must coincide.   □

Corollary. *A polynomial of degree $n>0$ has at most n distinct roots.*

EXERCISES

IX.1. Find the g.c.d. of the polynomials over **Q**:
$$X^5 - X^4 - 6X^3 - 2X^2 + 5X + 3, \quad X^3 - 3X - 2.$$
Also, find their g.c.d. over the field $\mathbf{F}_3$ if the coefficients are interpreted as congruence classes modulo 3.

IX.2. Show that $X^4 + 1$ is a prime polynomial over **Q**, but has divisors of degree 2 over the field defined in exercise VI.12.

IX.3. Let K be any field, and R a subring of $K[X]$ containing K. Prove that there exists a finite set of polynomials $P_1, P_2, \ldots, P_N$ in R such that R consists of all the polynomials in $P_1, P_2, \ldots, P_N$ with coefficients in K (Hint: call d the g.c.d. of the degrees of all polynomials in R, take $P_1, P_2, \ldots, P_m$ in R such that the g.c.d. of their degrees is d, and then apply the conclusion in exercise III.6).

Lemma. *Let G be a group of order m. If, for every divisor d of m, there are no more than d elements of G satisfying $x^d = 1$, G is cyclic.*

For every divisor d of m, call $\psi(d)$ the number of elements of order d in G; we have to prove that $\psi(m) > 0$. In any case, since every element of G has an order which divides m, we have

$$m = \sum_{d \mid m} \psi(d).$$

If, for some d, $\psi(d) > 0$, then G has an element of order d, and this generates a cyclic group G' of order d. As all the d elements of G' satisfy $x^d = 1$, our assumption on G implies that all the $\psi(d)$ elements of G of order d are in G'; by corollary 3 of theorem VIII.1, there are exactly $\varphi(d)$ such elements. Therefore, if $\psi(d)$ is not 0, it has the value $\varphi(d)$. Since $\sum \psi(d)$ has the value m, and $\sum \varphi(d)$ has the same value by corollary 4 of theorem VIII.1, this implies that $\psi(d) = \varphi(d)$ for all d; in particular, $\psi(m) = \varphi(m) > 0$.

Now we consider an arbitrary field K, and, denoting by $K^\times$ the multiplicative group of the non-zero elements of K, we consider the elements and subgroups of finite order of $K^\times$. If x is an element of $K^\times$ of order m, it satisfies $x^m = 1$, and $x^a = x^b$ if and only if $a \equiv b \pmod{m}$; traditionally, x is then called a *root of unity*, and more precisely a *primitive m^{th} root of unity*. For any n, an element x of K which satisfies $x^n = 1$ is a root of unity whose order divides n. In the field of complex numbers, the number

$$e^{2\pi i/m} = \cos\frac{2\pi}{m} + i\sin\frac{2\pi}{m}$$

is a primitive m^{th} root of unity; so is $e^{2\pi i a/m}$ for $(a,m) = 1$.

Theorem X.1. *If K is any field, every finite subgroup of $K^\times$ is cyclic.*

For every $n > 0$, an element of K satisfying $x^n = 1$ is a root of the polynomial $X^n - 1$; by the corollary of theorem IX.2, there are at most n such elements in K. Our theorem follows now at once from the lemma.

Corollary 1. *If K is a finite field, $K^\times$ is cyclic.*

Corollary 2. *If K is any field, and n an integer > 0, the elements of K satisfying $x^n = 1$ make up a cyclic group whose order divides n.*

It is clear that they make up a subgroup of $K^\times$; this is cyclic, by theorem X.1; if it is generated by x, the order of x, which is also the order of the group, must divide n.

Theorem X.2. *If p is any prime, there is an integer r prime to p such that $1, r, r^2, r^3, \ldots, r^{p-2}$, in some order, are respectively congruent to $1, 2, \ldots, p-1$ modulo p.*

This is only the traditional formulation for the fact that the congruence classes prime to p modulo p make up the multiplicative group $\mathbf{F}_p^\times$ of the field $\mathbf{F}_p$ of congruence classes modulo p, and that, by corollary 1 of theorem X.1, this must be cyclic; if $(r \bmod p)$ is a generator of that group, r has the property stated in theorem X.2.

If m is an integer > 1, the multiplicative group of congruence classes prime to m modulo m is not always cyclic (cf. e.g. exercises VII.7 and VII.8). It is cyclic if and only if there is an integer r prime to m such that $(r \bmod m)$ is of order $\varphi(m)$ in that group, i.e., if and only if the smallest integer $x > 0$ such that $r^x \equiv 1 \pmod{m}$ is $\varphi(m)$; when that is so, r is called a *primitive root modulo* m. Then, to every integer a prime to m, there is an integer x such that $r^x \equiv a \pmod{m}$; this integer x, which is determined only modulo $\varphi(m)$, is called the *index* of a and denoted by $\mathrm{ind}_r(a)$. By theorem VII.2, if r is a primitive root modulo m, the mapping

$$(a \bmod m) \to (\mathrm{ind}_r(a) \bmod \varphi(m))$$

is an isomorphism of the multiplicative group of congruence classes prime to m modulo m onto the additive group of all congruence classes modulo $\varphi(m)$. In particular, we have, for a and b prime to m:

$$\mathrm{ind}_r(ab) \equiv \mathrm{ind}_r(a) + \mathrm{ind}_r(b) \pmod{\varphi(m)}.$$

The analogy with logarithms is obvious.

X.1. If m is any integer >1, show that the number of primitive roots modulo m is either 0 or $\varphi(\varphi(m))$.

X.2. Find a primitive root r modulo 13; tabulate $\text{ind}_r(a)$ for $1 \leqslant a \leqslant 12$; use the table to find all primitive roots modulo 13, and to tabulate 5th and 29th powers modulo 13.

X.3. Use the existence of a primitive root modulo p, where p is a prime, to show that $1^n + 2^n + \cdots + (p-1)^n$ is congruent to 0 or to -1 modulo p according to the value of the integer $n \geqslant 0$.

X.4. Show that a primitive root modulo an integer $m > 1$ is also a primitive root modulo every divisor of m (Hint: use exercise V.6).

X.5. Using the binomial formula, prove by induction that, if p is an odd prime, then, for all $n \geqslant 0$:

$$(1+px)^{p^n} \equiv 1 + p^{n+1}x \pmod{p^{n+2}}$$

(cf. exercise VI.10). Hence show that, if r is a primitive root modulo p, it is a primitive root modulo p^n if and only if p^2 does not divide $r^{p-1}-1$, and that in any case either r or $r+p$ is a primitive root modulo p^n.

X.6. Find all the integers $m > 1$ such that there exists a primitive root modulo m (Hint: use exercises X.4, X.5, VII.7, VII.8, and the fact that if r is a primitive root modulo an odd integer m, then either r or $r+m$ is a primitive root modulo $2m$).

X.7. An integer $m > 0$ is called "square-free" if it has no divisor of the form n^2 where n is an integer > 1. For every $m > 0$, put $\mu(m) = (-1)^r$ if m is square-free and the product of r primes (with $r = 0$ if $m = 1$), and $\mu(m) = 0$ otherwise. Prove that $\mu(ab) = \mu(a)\mu(b)$ if a is prime to b; hence show that $\sum_{d|m} \mu(d)$ is 1 if $m = 1$, and 0 if $m > 1$ (Hint: write m as in exercise IV.4).

X.8. Let K be a field containing a primitive m^{th} root of unity x; for each divisor d of m, call $F_d(X)$ the product of the factors $X - x^a$ for $0 \leqslant a < m$, $(a, m) = \dfrac{m}{d}$. Show that F_d is of degree $\varphi(d)$ and prove the formula

$$X^m - 1 = \prod_{d \mid m} F_d(X);$$

hence, using exercise X.7, prove that

$$F_m(X) = \prod_{d \mid m} (X^{m/d} - 1)^{\mu(d)}.$$

X.9. K being as in exercise X.8, prove that the sum of all primitive m^{th} roots of unity in K is $\mu(m)$. State the special case of this result for $K = \mathbf{F}_p$, $m = p - 1$.

Now we will consider equations of the form $x^m = a$, in a field K (or occasionally in a ring); the case $a = 1$ has been discussed in § X. As the case $a = 0$ is trivial, we assume $a \neq 0$. If then, in the field K, x is a solution of $x^m = a$, an element x' of K is also a solution if and only if $(x'/x)^m = 1$. Therefore, if $x^m = a$ has a solution in K, it has as many solutions as K contains m^{th} roots of unity, i.e. roots of $X^m - 1$.

Here we are chiefly concerned with the field $\mathbf{F}_p$ of congruence classes modulo a prime p.

Theorem XI.1. *Let p be a prime, m an integer > 0, and a an integer prime to p; put $d = (m, p - 1)$. Then the congruence $x^m \equiv a \pmod{p}$ has either exactly d solutions modulo p or no solution; it has a solution if and only if the congruence $y^d \equiv a \pmod{p}$ has a solution; this is so if and only if $a^{(p-1)/d} \equiv 1 \pmod{p}$, and there are exactly $\dfrac{p-1}{d}$ such values of a modulo p.*

We use the fact that the group $\mathbf{F}_p^\times$ is cyclic, or, what amounts to the same, that there is a primitive root r modulo p (cf. § X). Put $a \equiv r^t$, $x \equiv r^u$ (mod p), i.e. $t = \mathrm{ind}_r(a)$, $u = \mathrm{ind}_r(x)$. Then the congruence $x^m \equiv a$ (mod p) is equivalent to $mu \equiv t$ (mod $p-1$), and our conclusions follow at once from theorem VI.2, provided we note that $t \equiv 0$ (mod d) is equivalent to $\dfrac{p-1}{d} t \equiv 0$ (mod $p-1$), i.e. to $a^{(p-1)/d} \equiv 1$ (mod p).

Take for instance the congruence $x^3 \equiv a$ (mod p), with a prime to p. For $p = 3$, this is equivalent to $x \equiv a$ (mod 3). Take the case where $p \equiv 1$ (mod 3); as this implies $p \neq 2$, p is also $\equiv 1$ (mod 2), hence of the form $6n+1$; we have $d = 3$, $\dfrac{p-1}{d} = 2n$; the congruence $x^3 \equiv a$ (mod p) has a solution if and only if a is congruent to one of the numbers $1, r^3, \ldots, r^{p-4}$ modulo p, and then, if x is one solution, the solutions are given by xr^{2nz} modulo p with $z = 0, 1, 2$. If $p \equiv 2$ (mod 3), in which case p is either 2 or of the form $6n-1$, the congruence $x^3 \equiv a$ (mod p) has one and only one solution for every a prime to p.

From now on, we consider only the case $m = 2$. Then $x^2 \equiv 1$ (mod p) has no other solution than 1 if $p = 2$, and has the two solutions ± 1 if $p > 2$.

Definition. If p is an odd prime, an integer a prime to p is called a quadratic residue or a quadratic non-residue modulo p according as the congruence $x^2 \equiv a$ (mod p) has a solution or not.

As no other case than $m = 2$ will occur, the word "quadratic" will occasionally be omitted; otherwise one speaks of "cubic residues" if $m = 3$, "biquadratic residues" if $m = 4$, etc.

Let p be an odd prime; put $p = 2n+1$, and let r be a primitive root modulo p. Then theorem XI.1 shows that there are n quadratic residues modulo p, viz., the classes of $1, r^2, \ldots, r^{2n-2}$, and the same number of non-residues, viz., the classes of $r, r^3, \ldots, r^{2n-1}$. If x is a solution of $x^2 \equiv a$ (mod p), this congruence has the two solutions $\pm x$, and no other, modulo p.

Theorem XI.2. *Let $p = 2n + 1$ be an odd prime, and a an integer prime to p. Then a^n is congruent either to $+1$ or to -1 modulo p; a is a quadratic residue or a non-residue modulo p according as $a^n \equiv +1$ (mod p) or $a^n \equiv -1$ (mod p).*

Put $b = a^n$; by Fermat's theorem (i.e. the corollary of theorem VIII.3), we have $b^2 \equiv 1$ (mod p), hence $b \equiv \pm 1$ (mod p). We can now apply theorem XI.1.

Corollary. *-1 is a quadratic residue or a non-residue modulo the odd prime p, according as $p \equiv 1$ or $p \equiv -1$ (mod 4).*

In fact, $(-1)^n$ is $+1$ or -1 according as n is even or odd.

EXERCISES

XI.1. If p is an odd prime divisor of $a^2 + b^2$, where a, b are integers, show that p must be congruent to 1 modulo 4 unless it divides both a and b.

XI.2. If p is an odd prime, and a is prime to p, show that the congruence $ax^2 + bx + c \equiv 0$ (mod p) has two solutions, one or none according as $b^2 - 4ac$ is a quadratic residue, 0 or a non-residue modulo p.

XI.3. If m, n are mutually prime integers > 0, and F is a polynomial with integral coefficients, show that the congruence $F(x) \equiv 0$ (mod mn) has a solution if and only if

$F(x) \equiv 0 \pmod{m}$ and $F(x) \equiv 0 \pmod{n}$ both have solutions (Hint: use exercises V.6 and VI.1).

XI.4. If p is an odd prime, $n > 0$, and a is prime to p, prove by induction on n that the congruence $x^2 \equiv a \pmod{p^n}$ has a solution if and only if a is a quadratic residue modulo p: show that, if x is a solution, there are no other solutions than $\pm x$ modulo p^n.

XI.5. Show that, if a is an odd integer, and $n > 2$, the congruence $x^2 \equiv a \pmod{2^n}$ has a solution if and only if $a \equiv 1 \pmod 8$ (Hint: use induction on n, observing that, if x is a solution, either x or $x + 2^{n-1}$ is a solution of $y^2 \equiv a \pmod{2^{n+1}}$). If x is a solution, find the other solutions.

XI.6. Use exercises XI.3,4,5 to give a criterion for the congruence $x^2 \equiv a \pmod{m}$ to have a solution when m is any integer > 1, and a is prime to m.

XI.7. If, for some $m > 1$ and some a prime to m, the congruence $x^2 \equiv a \pmod{m}$ has exactly n distinct solutions modulo m, show that there are exactly $\dfrac{\varphi(m)}{n}$ values of a, prime to m modulo m, for which this is so.

§ XII

Let p be an odd prime; put $p = 2n + 1$. Write G for the multiplicative group $\mathbf{F}_p^\times$ of congruence classes prime to p modulo p; this has a subgroup H of order 2 consisting of the classes $(\pm 1 \bmod p)$; we apply to G and H the definitions and lemma of § VIII. If x is an element of G, it belongs to one and only one coset xH; this consists of the two elements $(\pm x \bmod p)$; there are n such cosets, viz., the cosets $(\pm 1 \bmod p), (\pm 2 \bmod p), \ldots, (\pm n \bmod p)$. If, in each coset, we choose one element, and if we write these elements, in any order, as $u_1, \ldots, u_n$, this is known as "a set of representatives" for the cosets of H in G; then every integer prime to p is congruent to one and only one of the integers $\pm u_1, \ldots, \pm u_n$ modulo p. For the purposes of the next lemma, which is due to Gauss and known as Gauss' lemma, such a set $\{u_1, \ldots, u_n\}$ will be called a "Gaussian set" modulo p. The simplest such set is $\{1, 2, \ldots, n\}$.

Lemma. *Let $p = 2n + 1$ be an odd prime, and $\{u_1, \ldots, u_n\}$ a Gaussian set modulo p. Let a be an integer prime to p; for $1 \leqslant i \leqslant n$, let $e_i = \pm 1$ and i' be such that $au_i \equiv e_i u_{i'} \pmod{p}$. Then a is a quadratic residue modulo p or not according as the product $e_1 e_2 \ldots e_n$ is $+1$ or -1.*

In the n congruences $au_i \equiv e_i u_{i'} \pmod{p}$, no two values of i' can be the same, since otherwise this would give $au_i \equiv \pm au_k \pmod{p}$ for some $i \neq k$, hence $u_i \equiv \pm u_k \pmod{p}$, which contradicts the definition of a Gaussian set. Therefore, if we take the product of all these congruences, we get

$$a^n(u_1 u_2 \ldots u_n) \equiv (e_1 e_2 \ldots e_n) \cdot (u_1 u_2 \ldots u_n) \pmod{p}$$

and therefore

$$a^n \equiv e_1 e_2 \ldots e_n \pmod{p}$$

since all the u_i are prime to p. Our conclusion follows now from theorem XI.2.

Theorem XII.1. *Let p be an odd prime; then 2 is a quadratic residue modulo p if $p \equiv 1$ or $7 \pmod{8}$, and a non-residue if $p \equiv 3$ or $5 \pmod{8}$.*

Put $p = 2n + 1$, and apply Gauss' lemma to $a = 2$ and the Gaussian set $\{1, 2, \ldots, n\}$. If $n = 4m$ or $4m + 1$, e_i has the value $+1$ for $1 \leqslant i \leqslant 2m$, and -1 otherwise; then the product of the e_i is $(-1)^{n-2m} = (-1)^n$. If $n = 4m + 2$ or $4m + 3$, e_i is $+1$ for $1 \leqslant i \leqslant 2m + 1$, and -1 otherwise, and the product of the e_i is $(-1)^{n-2m-1} = (-1)^{n-1}$. A straightforward application of the lemma gives now the result stated above.

Definition. If p is an odd prime, and a an integer prime to p, we define $\left(\dfrac{a}{p}\right)$ to have the value $+1$ or -1 according as a is a quadratic residue or not modulo p; this is called the Legendre symbol.

When p is given, the symbol $\left(\dfrac{a}{p}\right)$ depends only upon the congruence class of a modulo p. Its definition implies that $\left(\dfrac{a^2}{p}\right)=1$ for all a prime to p.

If r is again a primitive root modulo p, and if $a\equiv r^x$ (mod p), i.e. $x=\mathrm{ind}_r(a)$, we have $\left(\dfrac{a}{p}\right)=(-1)^x$; here one should note that this does not depend upon the choice of x, since x is well defined modulo the even integer $p-1$. From the fundamental property of the index (cf. the last formula of § X), it follows that the Legendre symbol has the property

$$\left(\frac{ab}{p}\right)=\left(\frac{a}{p}\right)\cdot\left(\frac{b}{p}\right) \quad \text{for all } a,b \text{ prime to } p.$$

Theorems XI.2, its corollary, and theorem XII.1, give respectively:

$$a^{(p-1)/2}\equiv\left(\frac{a}{p}\right) \quad (\mathrm{mod}\,p), \quad \left(\frac{-1}{p}\right)=(-1)^{(p-1)/2},$$

$$\left(\frac{2}{p}\right)=(-1)^{(p^2-1)/8}$$

(as to the last formula, observe that $\dfrac{p^2-1}{8}$ is always an integer, even if $p\equiv1$ or 7 (mod 8) and odd if $p\equiv3$ or 5 (mod 8)).

The following theorem is known as "the quadratic reciprocity law":

Theorem XII.2. *If p and q are distinct odd primes, then*

$$\left(\frac{p}{q}\right) \cdot \left(\frac{q}{p}\right) = (-1)^{[(p-1)/2] \cdot [(q-1)/2]}$$

Put $p = 2n+1$, $q = 2m+1$. Apply Gauss' lemma to $a = q$ and to the "Gaussian set" $\{1, 2, \ldots, n\}$ modulo p. For $1 \leqslant x \leqslant n$, we have to write $qx \equiv e_x u \pmod{p}$ with $e_x = \pm 1$ and $1 \leqslant u \leqslant n$; this can be written as $qx = e_x u + py$, where e_x, u, y are uniquely determined by these conditions when x is given. In particular, e_x has the value -1 if and only if we have $qx = py - u$, or equivalently $py = qx + u$, with $1 \leqslant u \leqslant n$. This implies $y > 0$, and also

$$y \leqslant \frac{1}{p}(q+1)n < \frac{q+1}{2} = m+1.$$

In other words, we have $e_x = -1$ if and only if we can find y such that the pair (x, y) satisfies the conditions

$$1 \leqslant x \leqslant n, \quad 1 \leqslant y \leqslant m, \quad 1 \leqslant py - qx \leqslant n.$$

Consequently, if N is the number of such pairs (x, y), Gauss' lemma gives $\left(\dfrac{q}{p}\right) = (-1)^N$.

Similarly, $\left(\dfrac{p}{q}\right)$ has the value $(-1)^M$, if M is the number of pairs (x, y) satisfying

$$1 \leqslant x \leqslant n, \quad 1 \leqslant y \leqslant m, \quad 1 \leqslant qx - py \leqslant m.$$

As $qx - py$ cannot be 0 if x is prime to p, and in particular if $1 \leqslant x \leqslant n$, this shows that the left-hand side of the formula in our theorem has the value $(-1)^{M+N}$, where $M + N$ is the number of pairs (x, y) satisfying the conditions

$$1 \leqslant x \leqslant n, \quad 1 \leqslant y \leqslant m, \quad -n \leqslant qx - py \leqslant m.$$

Now let S be the number of pairs (x,y) satisfying

$$1 \leqslant x \leqslant n, \quad 1 \leqslant y \leqslant m, \quad qx - py < -n,$$

and let T be the number of pairs (x',y') satisfying

$$1 \leqslant x' \leqslant n, \quad 1 \leqslant y' \leqslant m, \quad qx' - py' > m.$$

Between those last two sets, there is a one-to-one correspondence defined by

$$x' = n + 1 - x, \quad y' = m + 1 - y;$$

in fact, when that is so, we have

$$qx' - py' - m = -(qx - py + n).$$

Therefore $S = T$. On the other hand, $M + N + S + T$ is the total number of pairs (x,y) with $1 \leqslant x \leqslant n$, $1 \leqslant y \leqslant m$, and is therefore equal to mn. Finally we have

$$\left(\frac{p}{q}\right) \cdot \left(\frac{q}{p}\right) = (-1)^{M+N} = (-1)^{M+N+S+T} = (-1)^{mn},$$

as was to be proved.

EXERCISES

XII.1. Let p be an odd prime; let $f(a)$ be a function, defined for a prime to p, taking no other values than ± 1, and such that

$$f(ab) = f(a)f(b); \quad f(a) = f(b) \text{ if } a \equiv b \pmod{p}.$$

Show that either $f(a) = 1$ for all a, or $f(a) = \left(\dfrac{a}{p}\right)$ for all a.

XII.2. If p is an odd divisor of $a^2 + 2b^2$, where a, b are integers, show that p is congruent to 1 or to 3 modulo 8 unless it divides both a and b.

XII.3. If p and q are primes such that $p = 2q + 1$ and $q \equiv 1$ (mod 4), show that 2 is a primitive root modulo p.

XII.4. Using only Gauss' lemma, find all the values of the prime $p > 3$ for which 3 is a quadratic residue.

XII.5. Let a be a non-zero integer. Prove that, if p and q are odd primes, not divisors of a, such that $p \equiv q$ (mod $4|a|$), then $\left(\dfrac{a}{p}\right) = \left(\dfrac{a}{q}\right)$ (Hint: write $a = \pm n^2 b$, where b is "square-free" (cf. exercise X.7); then apply the quadratic reciprocity law to every odd prime divisor of b and to p and q; also, apply theorem XII.1 if b is even, and the corollary of theorem XI.2 if $a < 0$).

We recall the concept of a complex number; this is a number of the form $x + iy$, where x, y are real numbers; i satisfies $i^2 = -1$, and the rules for addition and multiplication are the well-known ones. In particular, multiplication is given by

$$(x + iy)(x' + iy') = (xx' - yy') + i(yx' + xy').$$

For addition and multiplication, the set **C** of complex numbers makes up a unitary ring, with the unit element $1 = 1 + i \cdot 0$ (cf. § VI). If $a = x + iy$, one writes $\bar{a} = x - iy$, and calls $\bar{a}$ the imaginary conjugate of a; the imaginary conjugate of $\bar{a}$ is then a. The mapping $a \to \bar{a}$ is a one-to-one mapping of **C** onto itself, preserving the operations of addition and multiplication; it is thus an "automorphism" of **C**, i.e. an isomorphism of **C** onto itself.

We write $N(a) = a\bar{a}$, and call this the *norm* of a. By the multiplication rule, if $a = x + iy$, $N(a) = x^2 + y^2$; by the commutativity of multiplication, we have $N(ab) = N(a)N(b)$. The norm of a is 0 if and only if a is 0;

otherwise it is a real number >0. Consequently, for any $a = x + iy \neq 0$, we may put

$$a' = N(a)^{-1}\bar{a} = \frac{x}{N(a)} - i\frac{y}{N(a)},$$

and then $aa' = 1$, and, for every b, $a(a'b) = b$; conversely, if $az = b$, we have $a'(az) = a'b$, hence, by associativity, $z = a'b$. This shows that $\mathbf{C}$ is in fact a field. As usual, we associate, with the complex number $a = x + iy$, the point (x,y) in the plane; its Euclidean distance from the "origin" 0 is then $|a| = N(a)^{1/2}$; this is also called the "absolute value" of a.

For our purposes, we have to consider, rather than the field $\mathbf{C}$, the subset of $\mathbf{C}$ consisting of the complex numbers $x + iy$ where x,y are in $\mathbf{Z}$, i.e. ordinary integers. It can be verified at once that this is a ring; it is called the *Gaussian ring*, and its elements are called *Gaussian integers*. Clearly $a \to \bar{a}$ is an automorphism of that ring. If a is any Gaussian integer, $N(a) = a\bar{a}$ is a positive integer in $\mathbf{Z}$. Occasionally we also consider the numbers $x + iy$ where x,y are in $\mathbf{Q}$, i.e. rational numbers; just as above one sees that they make up a field (the "Gaussian field").

If a,b,x are Gaussian integers and $b = ax$, b is said to be a *multiple of a*; a is said to *divide b* or to be a *divisor of b*; when that is so, $N(a)$ divides $N(b)$. Every Gaussian integer divides its norm.

A divisor of 1 is called *a unit*; if $a = x + iy$ is a unit, $N(a)$ must divide 1 and so has to be 1; as x,y are integers, this implies that one of them is ± 1 and the other 0. Therefore the Gaussian units are ± 1, $\pm i$.

Two non-zero Gaussian integers a,b divide each other if and only if they differ only by a factor which is a unit, i.e. if $b = ea$ with $e = \pm 1$ or $\pm i$; then they are called *associates*. Among the four associates of a given integer

$a \neq 0$, there is one and only one, say $b = x + iy$, satisfying $x > 0$, $y \geqslant 0$; that one will be called *normalized*. For instance, among the associates $\pm 1 \pm i$ of $1 + i$, $1 + i$ and no other is normalized. Geometrically, the points in the plane, corresponding to the associates of a, are those deduced from a by a rotation around 0 by an angle $\frac{\pi n}{2}$ with $n = 0, 1, 2, 3$; the normalized one is the one which lies either on the positive real axis or in the "first quadrant".

A Gaussian integer of norm > 1 is called a *Gaussian prime* if it has no other divisor than the units and its own associates. It amounts to the same to say that q is a Gaussian prime if it is not 0 or a unit and has no divisor whose norm is > 1 and $< N(q)$. Ordinary integers which are prime in the previously defined sense (§ IV) will be called *rational primes* (or "ordinary primes"). If q is a Gaussian integer and $N(q)$ is a rational prime, then q is a Gaussian prime; as we shall see, the converse is not true. The associates of a Gaussian prime are Gaussian primes; as we have seen, there is one and only one of these which is "normalized" in the above defined sense. If q is any Gaussian prime, so is $\bar{q}$. If a is any Gaussian integer, not 0 nor a unit, then its divisor of smallest norm > 1 must be a Gaussian prime.

Gauss, who introduced Gaussian integers into number-theory, found that they can be (essentially uniquely) factorized into Gaussian primes, in close analogy with ordinary integers. This will now be proved; the method is the same as in §§ II, III, IV (and as in § IX). We first prove a lemma, similar to those in § II and §IX.

Lemma. *Let a, b be Gaussian integers, with $b \neq 0$. Then there is a multiple bq of b such that*

$$N(a - bq) \leqslant \tfrac{1}{2} N(b).$$

For any real number t, there is a largest integer $m \leqslant t$, and we have $m \leqslant t < m+1$; by the nearest integer m' to t, we will understand either m or $m+1$ according as $t-m$ is $\leqslant m+1-t$ or not; then we have $|t-m'| \leqslant \frac{1}{2}$. Now let $z = x + iy$ be any complex number; call m the nearest integer to x, n the nearest integer to y, and put $q = m + in$. Then q is a Gaussian integer, and we have

$$N(z-q) = (x-m)^2 + (y-n)^2 \leqslant \tfrac{1}{2}.$$

Apply this to $z = \dfrac{a}{b}$, where a, b are the integers in the lemma. The Gaussian integer q defined by the above construction has then the required property.

Theorem XIII.1. *Let $\mathfrak{M}$ be a non-empty set of Gaussian integers, closed under addition and such that, if a is in $\mathfrak{M}$, all multiples of a are in $\mathfrak{M}$. Then $\mathfrak{M}$ consists of all the multiples of some Gaussian integer d, uniquely determined up to a unit factor.*

If $\mathfrak{M} = \{0\}$, the theorem is true with $d = 0$. If not, take in $\mathfrak{M}$ an element d of smallest norm > 0. If a is in $\mathfrak{M}$, we can apply the lemma and write $a = dq + r$ with $N(r) \leqslant \frac{1}{2} N(d)$. Then $r = a - dq$ is in $\mathfrak{M}$, which contradicts the definition of d unless $r = 0$, $a = dq$. As to unicity, if d' has the same property as d, d and d' must be multiples of each other, so that d' is an associate of d.

Just as in § II (and in § IX), we can now apply theorem XIII.1 to the set of all linear combinations $ax + by + \cdots + cz$, where $a, b, \ldots, c$ are given Gaussian integers and $x, y, \ldots, z$ are arbitrary Gaussian integers, and so define the g.c.d. $(a, b, \ldots, c)$; this will be uniquely determined if we prescribe that it should be "normalized" (in the sense defined above). If it is 1, we say that $a, b, \ldots, c$ are mutually relatively prime. We can now repeat the contents of §§ III and IV, except that the proof of theorem IV.2 was by induction on the integer a in that theorem, while now one has to use induction on $N(a)$. The conclusion is:

Theorem XIII.2. *Every non-zero Gaussian integer can be written "essentially uniquely" as a product of a unit and of Gaussian primes.*

Here the words "essentially uniquely" have the following meaning. Let

$$a = e q_1 q_2 \ldots q_r = e' q_1' q_2' \ldots q_s'$$

be two products of the required type for some $a \neq 0$, where e, e' are units and the q_j and q_k' are Gaussian primes. Then the theorem should be understood to say that $r = s$ and that the q_k' can be reordered so that q_j' is an associate of q_j for $1 \leqslant j \leqslant r$; if a is a unit, $r = 0$. If one prescribes that the prime factors of a should be "normalized", then the product is uniquely determined up to the order of the factors.

Ordinary integers are also Gaussian integers; in order to obtain their decomposition into Gaussian primes, it is enough to do this for "ordinary" primes.

Theorem XIII.3. *Let p be an odd rational prime. Then it is either a Gaussian prime or the norm of a Gaussian prime q; in the latter case $p = q\bar{q}$, q and $\bar{q}$ are not associates, and p has no other Gaussian prime divisor than q, $\bar{q}$ and their associates.*

Put $p = eq_1 q_2 \ldots q_r$ as in theorem XIII.2. Taking the norm, we find that p^2 is the product of the $N(q_j)$. If one of the $N(q_j)$ is p^2, then $r = 1$, $p = eq_j$, and p itself is a Gaussian prime. Otherwise each $N(q_j)$ is p, and we can write $p = N(q) = q\bar{q}$, with q a Gaussian prime; $\bar{q}$ is then also a Gaussian prime. Put $q = x + iy$; if $\bar{q}$ was an associate of q, it would be $\pm q$ or $\pm iq$; this gives either $y = 0$, $p = x^2$, or $x = 0$, $p = y^2$, or $y = \pm x$, $p = 2x^2$; this cannot be, since p is an odd prime.

As to $p = 2$, its decomposition is given by

$$2 = N(1 + i) = (1 + i)(1 - i) = i^3(1 + i)^2;$$

this has the only normalized prime factor $1 + i$.

Theorem XIII.4. *Let p be an odd rational prime. Then p is a Gaussian prime or the norm of a Gaussian prime according as it is congruent to 3 or to 1 modulo 4.*

If it is the norm of $q = x + iy$, we have $p = x^2 + y^2$, where one of the integers x, y must be odd and the other even. Then one of the squares x^2, y^2 is congruent to 1 and the other to 0 modulo 4, so that $p \equiv 1 \pmod{4}$. Conversely, if this is so, the corollary of theorem XI.2 shows that -1 is a quadratic residue modulo p, so that there is x such that $x^2 + 1$ is a multiple of p. As $x^2 + 1 = (x + i)(x - i)$, this, if p were a Gaussian prime, would imply that p divides either $x + i$ or $x - i$. Obviously this cannot be so.

Corollary 1. *Every Gaussian prime is either $\pm 1 \pm i$, or an associate of a rational prime congruent to 3 modulo 4, or else its norm is a rational prime congruent to 1 modulo 4.*

In fact, every Gaussian prime q must divide some prime rational factor p of its norm $q\bar{q}$; applying theorem

XIII.4 if p is odd, and the above remarks if $p=2$, we get our conclusion.

Corollary 2. *A rational prime can be written as a sum of two squares if and only if it is 2 or congruent to 1 modulo 4.*

In fact, if $p = x^2 + y^2$, p cannot be a Gaussian prime, since it has the divisors $x \pm iy$.

It is worthwhile pointing out that this is a statement concerning rational integers which has been proved using a larger ring, viz., the Gaussian integers.

EXERCISES

XIII.1. If a positive integer is written as $n^2 a$ where a is >1 and square-free (cf. exercise X.7), show that it can be written as a sum of two squares if and only if every odd prime divisor of a is $\equiv 1 \pmod 4$. If that is so, and a has r prime divisors, find the number of representations of a as a sum of two squares.

XIII.2. If an integer is the sum of two relatively prime squares, show that the same is true of every divisor of that integer.

XIII.3. Using the representation of complex numbers by points in the plane, show that, if z is any complex number, there is a Gaussian integer q whose distance to z is $\leqslant \dfrac{\sqrt{2}}{2}$; show that, among all Gaussian integers, there is at least one whose distance to z is smallest, and that there are no more than four with that property (Hint: cf. the proof of the lemma in § XIII).

XIII.4. The congruence relation being defined for Gaussian integers in the same way as for ordinary integers (cf. § V), call $f(m)$, for any Gaussian integer $m \neq 0$, the number of distinct Gaussian congruence classes modulo

m; show that $f(mn) = f(m)f(n)$ for any two non-zero Gaussian integers m, n (Hint: take representatives x_i, with $1 \leqslant i \leqslant f(m)$, for the classes modulo m, representatives y_j, with $1 \leqslant j \leqslant f(n)$, for the classes modulo n, and show that the $x_i + my_j$ are representatives of the classes modulo mn).

XIII.5. Use exercise XIII.4 to show that $f(m) = m\bar{m}$ for every m (Hint: apply exercise XIII.4 to m and to $n = \bar{m}$).

XIII.6. Show that, if m is a Gaussian integer of norm > 1, the Gaussian congruence classes modulo m make up a field if and only if m is a Gaussian prime. Show that, if $N(m)$ is a rational prime, each Gaussian integer is congruent to some rational integer modulo m.

XIII.7. If $\omega = -\dfrac{1}{2} + i\dfrac{\sqrt{3}}{2}$, show that the complex numbers $x + y\omega$, where x and y are ordinary integers, make up a ring R, whose units are $\pm 1, \pm \omega, \pm \omega^2$. Prove that, if z is any complex number, there is an element q of the ring R such that $N(z - q) \leqslant \frac{1}{3}$ (Hint: cf. exercise XIII.3). Hence prove the analogue of theorem XIII.1 for the ring R, and prove for that ring a unique factorization theorem.

XIII.8. Use exercise XIII.7 to show that a rational prime > 3 can be written as $x^2 + xy + y^2$, with integers x, y, if and only if it is $\equiv 1 \pmod 3$.

INDEX

Other books by André Weil

Basic Number Theory
Third Edition

"...this book consists of the body of ideas usually associated with the name of Class Field Theory, both local and global ... a coherent treatment of the known "basic" results on which our knowledge of number theory today is founded. It may well be regarded by latter historians as one of the "classics" in this field, compared perhaps to Hecke's "Theorie der algebraischen Zahlen" ... its emphasis lies on equipping the reader with the fundamental tools and a general "analytical outlook..."

Bulletin of the London Mathematical Society

1974/xviii, 325 pp./Cloth
(Grundlehren der mathematischen Wissenschaften, Volume 144)
ISBN 0-387-06935-6

Elliptic Functions According to Eisenstein and Kronecker

This book presents in two parts a modern, never-before-published treatment of Eisenstein's elementary but far-reaching methods in the theory of elliptic functions and its further development by Kronecker. Eisenstein's method, detailed here for the first time since his original paper of 1847, leads quickly and elegantly to all the classical theorems in the theory, as well as to results of considerable importance in modern number theory.

A systematic explanation of the topics connected with Kronecker's celebrated "limit formula" and its application is then presented in Part II. It also provides an introduction to the fundamental importance of much of Hecke's work whose contributions have also become an integral part of twentieth century mathematicians.

1976/xii, 92 pp./Cloth
(Ergebnisse der Mathematik und ihrer Grenzgebiete, Volume 88)
ISBN 0-387-07422-8

Œuvres Scientifiques
Collected Papers (1926-1978)
In Three Volumes

The entire mathematical work of André Weil, exclusive of his books is now available in this three-volume edition. Reflecting his enormous range of interest and dramatizing the decisive impact his work has had on many areas of contemporary mathematics, this collection contains both published and previously unpublished or otherwise inaccessible papers, including Weil's own comprehensive commentary on his work over the past 50 years. For everyone interested in mathematics, this essential working guide will remain for years to come *the* basic reference on one of this century's most important mathematicians.

1978/3 vols. approx. 500 pp. each/Cloth
(volumes not sold separately)
ISBN 0-387-90330-5

Dirichlet Series and Automorphic Forms
Lezioni Fermiane

1971/v, 164 pp./Paper
(Lecture Notes in Mathematics, Volume 189)
ISBN 0-387-05382-4